CONTENTS

The Dark Ages	5
The Arab civilisation	7
The Middle Ages	12
The Renaissance	16
Leonardo da Vinci	20
The break with the past begins	24
The Copernican system	28
The Counter-Reformation	33
Tycho Brahe and Kepler	35
Galileo	41

Ptolemy – a mediaeval representation

THE DARK AGES

When the barbarian tribes overran the Roman Empire and in AD 410 succeeded in capturing Rome itself, one institution alone stood firm in the midst of the general destruction – the Christian Church. The Church saw its task as that of trying to civilise the barbarian conquerors and of converting them to Christianity – a formidable undertaking in an age of violence when men and women were mainly concerned with keeping themselves alive. As the Roman Empire went down into darkness most forms of learning perished, including Greek science, which had never been of much account to the Romans for they had never possessed the Greek spirit of curiosity and enquiry. The Romans, it has been said, travelled with the aid of milestones; the Greeks travelled by the stars. It is not far wrong to say that between about AD 500 and 1000 nobody in western Europe thought much about anything other than self-preservation in this world and in the world to come.

The Eastern Empire, with its capital at Constantinople, showed few signs of being a centre from which learned men could develop the ideas of Aristotle and Ptolemy. The rulers of the Eastern Empire were preoccupied with theology and politics. They regarded the state as the embodiment of Christianity, much as today Russians might say that the USSR embodied the teachings of Karl Marx. All intellectual discussions tended to become theological and since the Greeks had been pagan there was little reason to show interest in their science. The Greek language was kept alive in Constantinople, and Greek manuscripts were highly prized, but more as relics than as stepping-stones to new ideas.

It was only by a long and very roundabout journey that Greek science returned to western Europe. The story starts in Alexandria, where between about 300 BC and its capture by the Moslems in AD 642 there lived a curious mixture of races – Egyptians, Phoenicians, Chaldaeans, Persians, Jews and Greeks – bound together by a common interest in Greek language and ideas and forming a vigorous centre of learning. They lived a comparatively untroubled life until the seventh

THE DARK AGES

century when, under the influence of Mahomet, the Arab conquests began and Moslems swiftly overran Asia Minor, Persia, North Africa and Spain. The important fact about these conquerors is that they respected the intellectual life of the peoples whose lands they invaded. Islam, at the start, was a tolerant religion, and although Alexandria ceased to have its old importance, the Greek ideas studied there were readily accepted by the Moslem invaders. The great age of Arab civilisation had begun.

Baghdad between AD 772 and 922

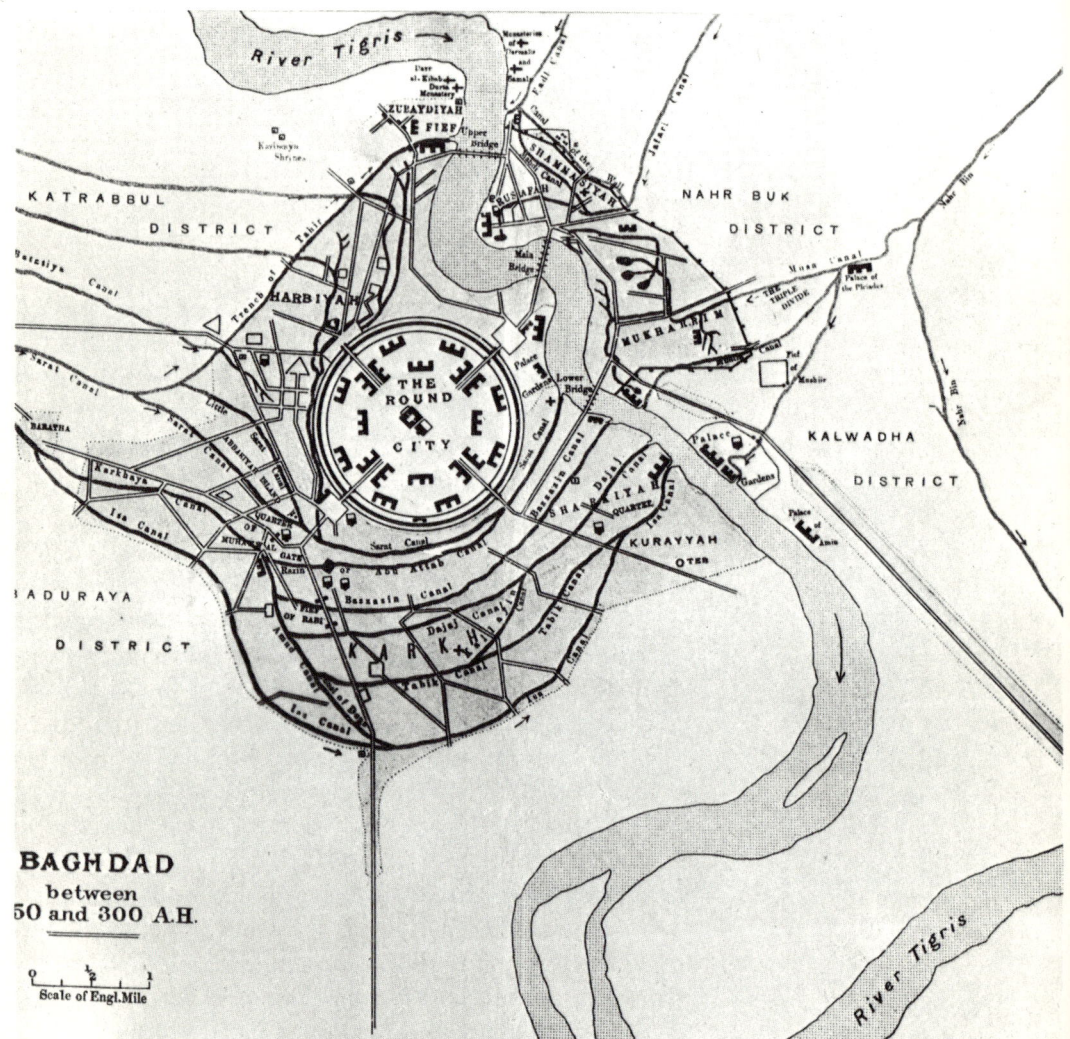

BAGHDAD between 50 and 300 A.H.

THE ARAB CIVILISATION

When we speak of Arabs we do not mean only the people living in Arabia; in fact most of the new discoveries were not made by Arabs in the strict sense of the term. The word is conveniently used to cover all Arabic-speaking Moslem peoples wherever they lived between Mesopotamia and Spain. The first centre of Arab civilisation was the splendid new city of Baghdad with its well-paved and lamp-lit streets, a strange contrast to European squalor. Under Caliphs like Haroun-al-Rashid the city became a centre for the study of mathematics, astronomy and medicine. By AD 850 the Arabs possessed translations of Aristotle, Euclid, Ptolemy and of the medical works of Galen. They had always been interested in astronomy, for their great trading caravans tended to travel by night to avoid the intense mid-day heat, and the stars were a necessary guide. In 829 an observatory was built in Baghdad and the astrolabe, an instrument something between a sextant and a computer, was invented to help in desert navigation. Substantial contributions were also made in medicine and chemistry, even though

A Syro-Egyptian astrolabe, c. late ninth century AD

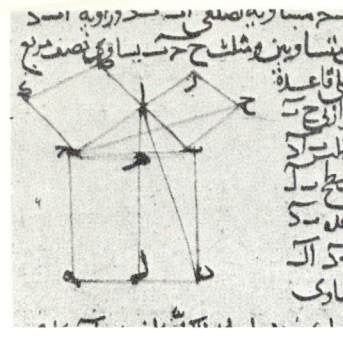
Arabic mathematics

the formula for an *elixir vitae*, or cure for all ills, and a method of transmuting metals into gold somewhat naturally eluded them. Arab merchants sailed to India, where they learned algebra from the Hindus and whence they brought back the so-called Arab numerals, to which they themselves added the symbol for zero. (A moment's thought of how complicated it must have been to divide, say, MDCCCXIV by DCLXVII shows the benefit brought to mathematics by the new figures.) Sailing still further east to Canton, they learned the art of paper-making.

Although the golden age of Baghdad lasted only until about AD 1000, a new centre of Arab intellectual activity arose to take its place at Cordoba in Spain. Despite constant fighting between Moslems and Christians in Spain, warfare did not then involve whole nations as it does today. There was always a steady flow of scholars across frontiers, and around 1150 a

The Mosque at Cordoba

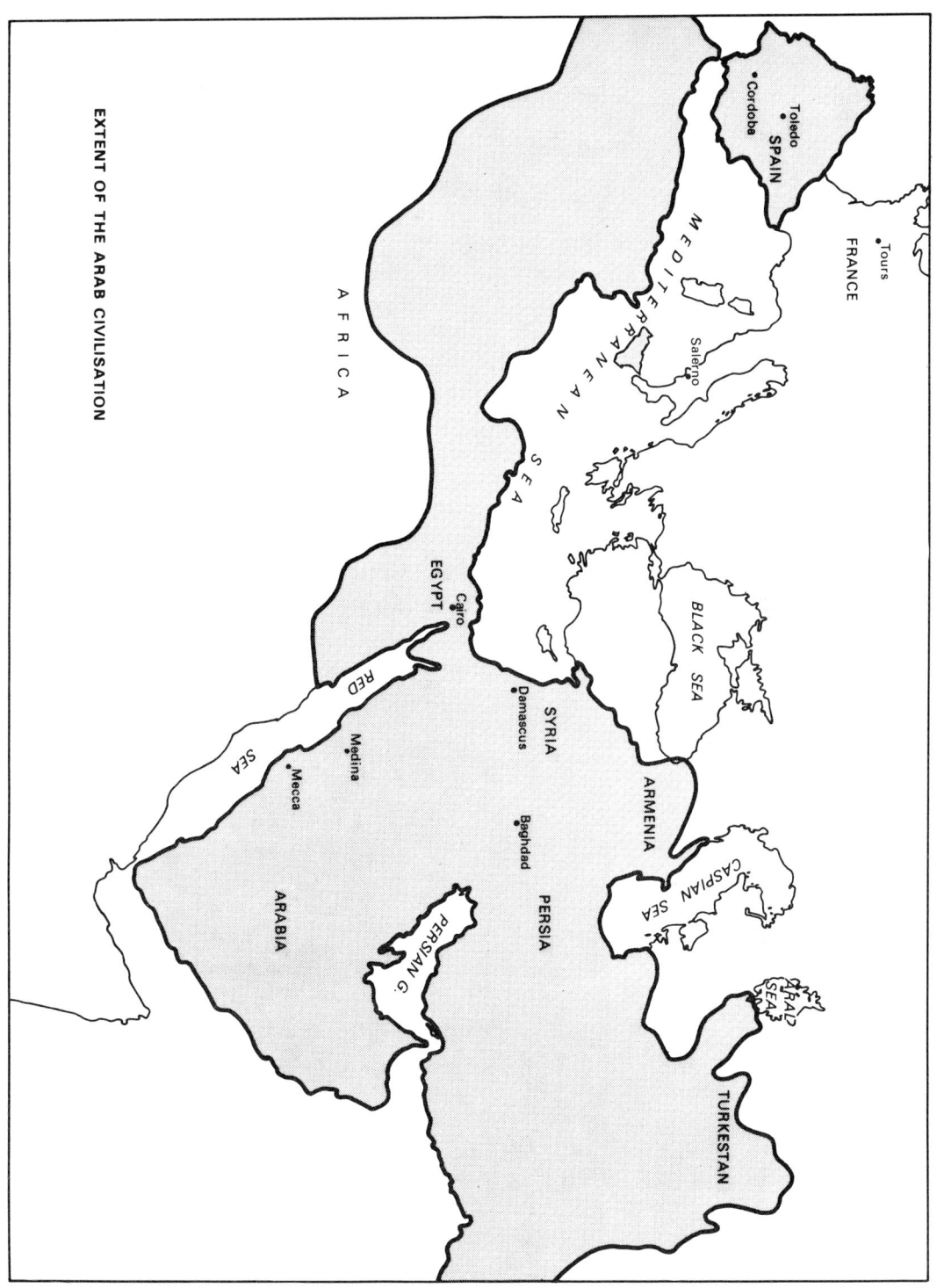

THE ARAB CIVILISATION

scholar called Gerard of Cremona set up in Toledo what was almost a translating factory. From Toledo translations of Greek works began to be taken across the Pyrenees into France. A similar traffic of ideas existed between Spain and Italy, for Sicily was in Arab hands and it was an easy step to the mainland. The immediate result was the setting up at Salerno, near Naples, of the first medical school.

From the start of the twelfth century the political power of Islam began to wane. In an attempt to preserve it, Moslem rulers became less tolerant and more ready to condemn scientific discussion as likely to corrupt the faith. Fortunately this decline of Arab civilisation coincided with a great revival of learning in western Europe stemming from the foundation between 1100 and 1300 of universities at Paris, Bologna, Oxford, Cambridge and many other cities, which quickly became famous. More stable government, combined with increasing wealth and greater security from violence, created the conditions needed for reviving the intellectual leadership of the West.

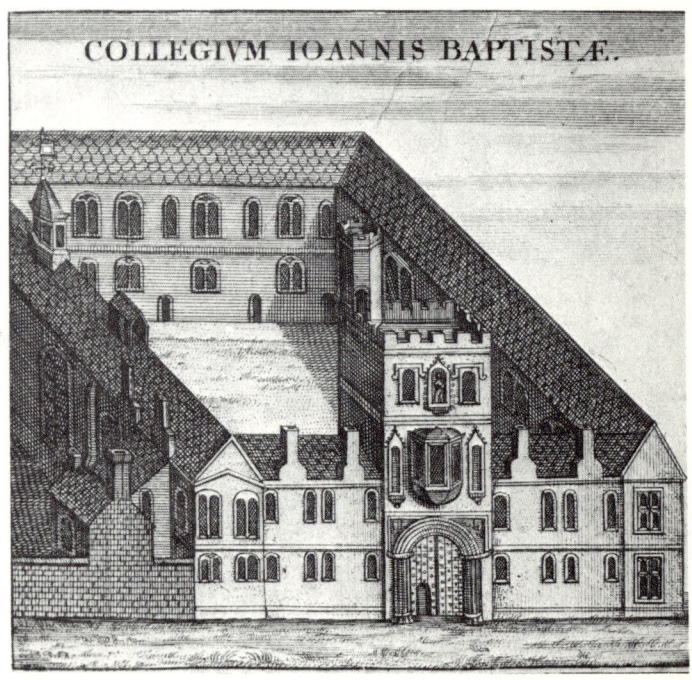

St John Baptist College at Oxford

UNIVERSITIES FOUNDED DURING THE LATER MIDDLE AGES

THE MIDDLE AGES

It is difficult to appreciate just how ignorant learned men in Europe had been round about AD 1000. Compared with the Arab world Europe was a backwater. The Royal Library in Paris had some 2,000 books at a time when Cordoba had a library with half a million volumes. Gradually the darkness began to be dispelled and in this task the Church played a crucial part. In the Middle Ages the Church provided the institutions where philosophy was taught and it was here that the spirit of free enquiry was fostered. The great thinkers and teachers of the twelfth and thirteenth centuries, the Schoolmen as they were called, were all churchmen, for they alone could read and write. 'Clerk' and 'cleric' are basically the same word. It was only later, at the time of the Reformation, that the spirit of intolerance tried to stamp out free thought.

Thanks to earlier contacts with Arab scholars all the works of Aristotle had been recovered between 1200 and 1250. In his writings the Schoolmen came upon an intellectual system beyond their previous experience. It was the task of a Dominican friar, St Thomas Aquinas (1227–74), to reconcile Christianity with the ideas of Aristotle: to make it possible for men to hold fast to the revealed truths of Christianity and the truths of human reason as expounded by Aristotle. These, according to Aquinas in his *Summa Theologica,* could not be at variance since both flowed from God. The importance of this work in the history of science is that for the next three centuries Greek thought was placed on almost the same pedestal as Christianity. Neither could be questioned.

However, one man did question rather more than others, a man born out of due time, the Franciscan friar, Roger Bacon (c. 1210-92). In spirit he was akin to the Arabs before him and to the men of the Renaissance who were to follow. He it was who, alone among his contemporaries, saw that experimental methods gave certainty to a scientific theory. 'There are two methods of investigation,' he wrote, 'through argument and through experiment. Argument does not suffice, but experiment does.' His own experiments lay mainly

THE MIDDLE AGES

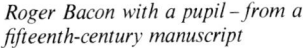

Roger Bacon with a pupil – from a fifteenth-century manuscript

with optics and with gunpowder, though he never progressed very far, probably because of the suspicion which his labours aroused in the minds of his superiors. For two long spells he was forbidden to write or teach, and for a time he was imprisoned – unusually harsh treatment for his time. Bacon deserves to be remembered for his glimpse of the true scientific method which would not be generally accepted until three centuries after his death.

The achievement of the Middle Ages was to recover and preserve for the use of later ages the science of classical times. This was both good and bad. It was valuable to know the point which Greek thinking had reached: it was disastrous when that thinking became embalmed and for too long placed above criticism. Despite this there were signs of progress. Mathematics was developed; the astrologers who studied the stars to determine their 'influence' on the careers and characters of men unwittingly prepared the way for Copernicus; the alchemists vainly searching for gold stumbled upon new facts about metals and gases; Albertus Magnus of Cologne (1206-80) known as 'the bishop with the boots' because he tramped all over his diocese, was the first naturalist to study insects since Aristotle and his work gave rise to later books about animals,

The Alchemist – a woodcut by Breughel

herbs and precious stones. In addition there was some degree of technical progress: the three-field system produced a rotation of crops; farming also benefited from the invention of a heavy wheeled plough and a new method of harnessing a horse to it; much land was reclaimed in Flanders and east Germany; and, above all, the superb Gothic cathedrals and parish churches of western Europe were not only a monument to religious faith but also a reminder of remarkable building techniques.

Yet when all is said and done the entire climate of

JANUARY PLOUGHING.

THE MIDDLE AGES

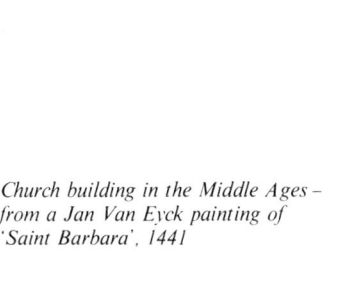

Church building in the Middle Ages – from a Jan Van Eyck painting of 'Saint Barbara', 1441

opinion was opposed to scientific advance. The Age of Faith was totally unaware of its need for science. If God was in his heaven all must be right with the world. Even medical knowledge received scant encouragement, for it was better to die well than to live badly. Technical improvements were held up by craft guilds which imposed secrecy on all members of the 'mystery' in order to prevent a man's livelihood being stolen by outsiders. Not until the Church lost its hold and a new view of man's place in the universe arose would the climate be right for the scientific revolution of the seventeenth century. In the history of science the Renaissance is the period in which men's minds were freed from the past and the possibility of a new world was born.

THE RENAISSANCE

At the outset the Renaissance was a literary movement, the scholarly study of long-lost but now recovered Latin and Greek texts. A study of the classics led to a renewed interest in such Roman and Greek architecture and sculpture as had survived the Dark Ages. This study helped to inspire the marvellous outburst of art which began in Florence in the fourteenth century and reached its height in the works of Leonardo, Michelangelo, Raphael and Titian in the first quarter of the sixteenth century. The early Renaissance was a time of

Renaissance Florence

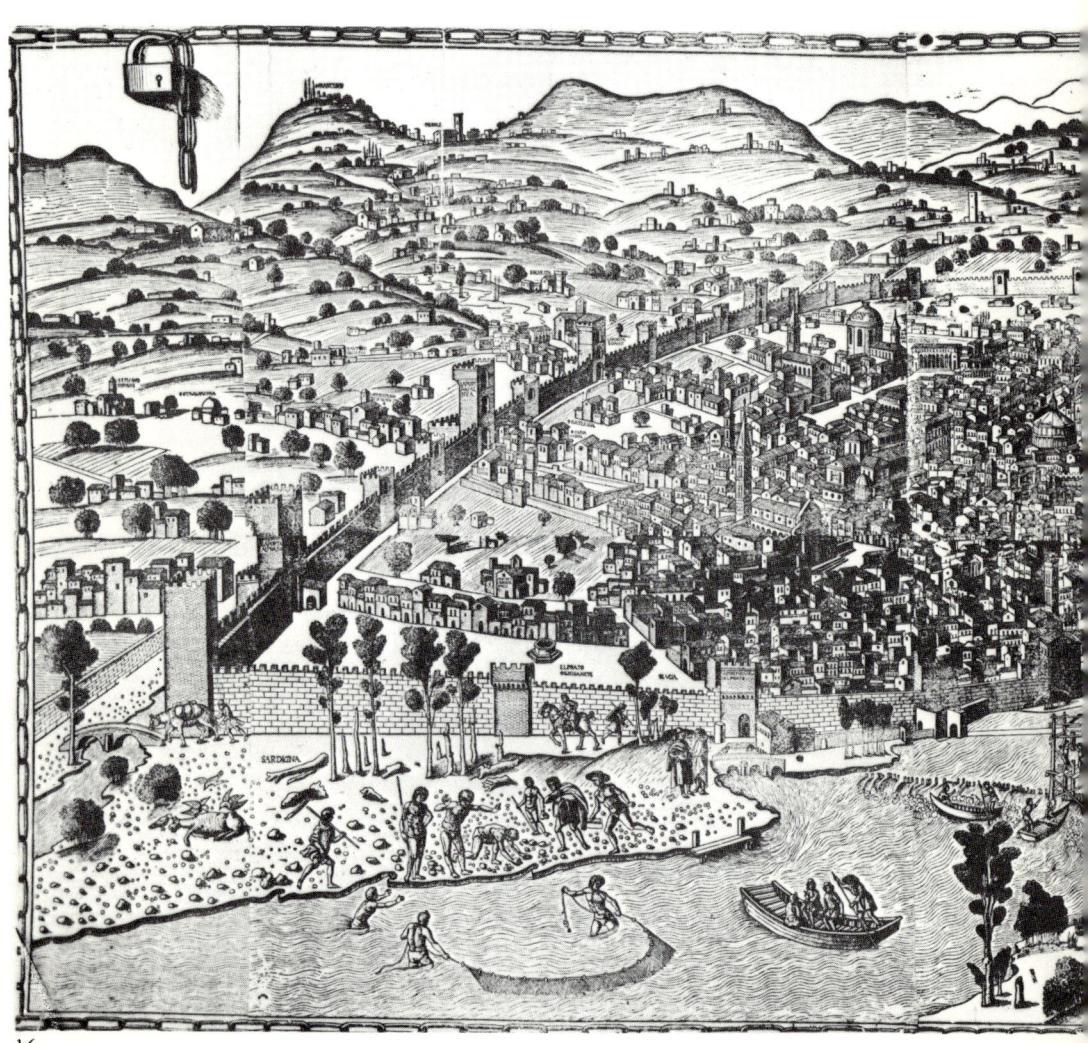

THE RENAISSANCE

great intellectual excitement among scholars and artists, who were convinced that they were living in a new age which promised to be even more glorious than the great days of Greece and Rome. It was not, however, a backward-looking age: the past provided inspiration, but the creations of the great artists were revolutionary in their ideas and technical achievement. It would be equally wrong to think of the Renaissance as belonging solely to Italy. Ideas could now spread more quickly, partly because of greater ease

THE RENAISSANCE

of communication in travel but much more because of the supremely important invention of printing in mid-fifteenth-century Germany. The laborious task of copying books by hand had prevented any rapid spread of knowledge. Without printing significant scientific advances would have been virtually impossible.

As far as science is concerned the most important aspect of the Renaissance was the change in men's attitude toward authority, especially the authority of the Church. Men did not cease to be Christian, but the hold of churchmen over the minds of those prepared to use their brains fearlessly was slowly lessening. The pilgrims who went to Canterbury in Chaucer's great

An early printing press

The Friar – from the Ellesmere manuscript drawings of The Canterbury Tales

poem, like the Nun, the Friar or the Pardoner, are close portraits of those whom he must have seen and mocked. The world was not necessarily evil and ugly as the Church had too often taught; it was a place of beauty with its secrets waiting to be revealed. Artists dissected corpses at night in order to study anatomy; the figures which they then painted were no longer heavily draped but, unashamedly naked, showed the beauty of the human body. The same exciting search for new knowledge inspired the men who set out from Portugal and Spain in tiny ships to discover 'the Indies', and instead found America and the sea route to the Far East. Something of the same spirit drove Shakespeare, greatest of Renaissance writers, to show in his plays the height and depth of human nature. More obviously this search for truth inspired men to study afresh the heavens and the movements of the stars, leading them to the irresistible conclusion that what Ptolemy and Aristotle had written was not the last but only the first word on astronomy.

When studying the story of Renaissance science it should be remembered that no exact dates can be given for the start and finish of such a movement. By 'Renaissance' we here mean the new intellectual approach which began in Italy in the fourteenth century, spread rapidly over most of western Europe and prepared the way for the true start of modern science in the seventeenth century. The Renaissance was the dawn of the new day and not the day itself. Thus it will not be surprising to find that many of the great men, despite their revolutionary ideas, were often still partly mediaeval in outlook. However, one giant figure is an exception to this rule, Leonardo da Vinci.

LEONARDO DA VINCI

By any standard of judgment Leonardo da Vinci must be regarded as one of the greatest intellects in the history of mankind. In his experimental approach, his acute power of observation and his profound thinking this man might have advanced science at one step to the point which it only reached a century after his death. Unfortunately, for reasons which we shall see later, his work was never published and his speculations remained unknown to all save his close friends.

Leonardo was born at Vinci near Florence in 1452. At fourteen he was apprenticed to the Florentine artist Verrocchio, until by 1472 his genius as a painter was apparent. Already his contemporaries realised that Leonardo was in a class apart from the rest of mankind. Striking good looks and great charm of manner concerned them less than the power of his mind and his force of character. For ten years he worked in Florence as both painter and sculptor and then entered the service of the Duke of Milan as a civil engineer as well as an artist. Here he stayed till 1499. The last twenty years of his life were spent moving between Venice, Florence and Rome until in 1516 he accepted an invitation from Francis I to settle at Amboise, where he remained until his death in 1519.

The first thing which must strike anyone about Leonardo is the quite extraordinary range of his abilities. In a famous letter offering his services to the Duke of Milan he spoke of his work not only as a painter and sculptor but also of his skill as an inventor, military engineer, poet, musician and town planner. He was also a physicist, biologist and philosopher, and in each of his many rôles he was supremely gifted.

His second great quality was boundless curiosity: there seemed to be nothing which did not excite his interest. It was probably his painting which first drew Leonardo to science and led him to make accurate observations of men and beasts, of plant life and rock formations. Although this is not the place to speak of Leonardo as an artist, it should never be forgotten that this far-reaching thinker was also the painter of 'The Last Supper' and of the 'Mona Lisa', perhaps the

Studies of skulls by Leonardo da Vinci

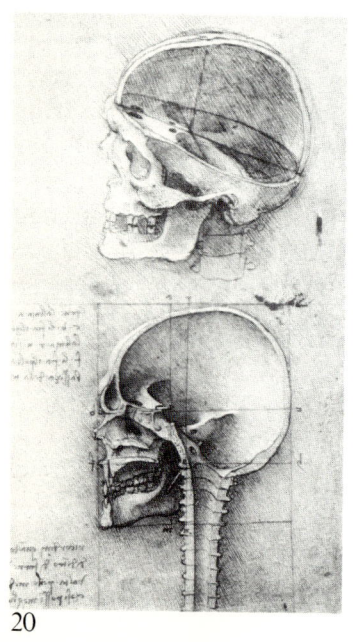

| LEONARDO DA VINCI |

Sketch of a blackberry by Leonardo da Vinci

two most familiar paintings in the world. Thus as a painter he came to study the laws of optics and the structure of the eye. He never ceased to be fascinated by all forms of movement – the human body in action (which led him secretly to dissect more than ten human corpses) or the minute movements of the muscles; the tangled forms of the growth of a blackberry or the exact movement of a bird's wing in flight. As he grew older and as his work as an engineer demanded it, he became increasingly preoccupied with water, its power,

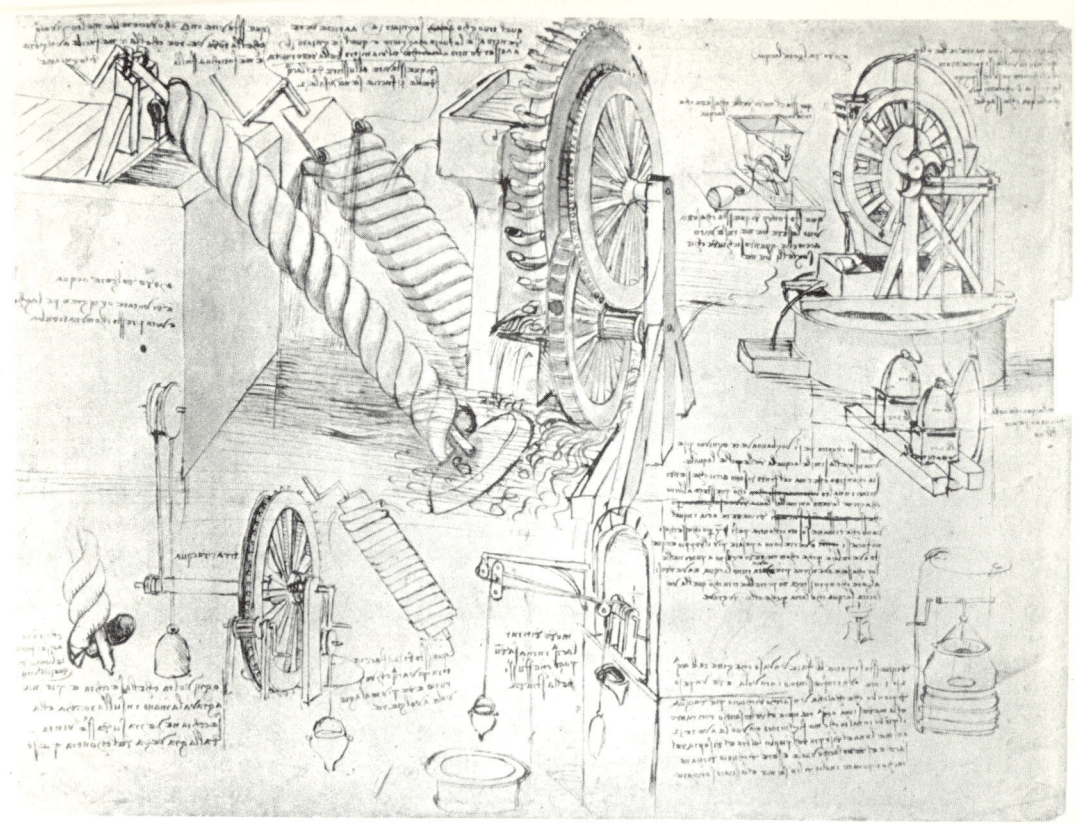

Archimedean screw – a page from a Leonardo notebook

with the fantastically complicated patterns of whirlpools, storms and tempests. Aristotle had written learnedly in a theoretical way about many of the things which interested Leonardo, but this was of little value compared with a direct study of people and things as they really are. Experiment must be the supreme test of any scientific theory.

It is uncertain whether Leonardo intended to publish his discoveries and speculations; certainly he died without doing so. Instead he left behind him a wonderful series of notebooks in which he confided his thoughts and the details of his experiments. He was almost certainly left-handed and his notes were written from right to left, with each letter inverted in mirror style. There is very little punctuation but a wealth of abbreviations. Hence the extreme difficulty of de-

LEONARDO DA VINCI

ciphering his notebooks, a task which was not seriously undertaken until the mid-nineteenth century. Each page is quite small and is covered not only with Leonardo's strange script but also with drawings of exquisite beauty and mastery of detail. It is the fact that all Leonardo's thought went into those notebooks which accounts for the comparatively small influence which he had in his lifetime, but today it is apparent how immense was the range of that thought. He quickly rejected the possibility of the idea of 'perpetual motion'; he regarded the lever as the primary machine, with all other machines as modifications of it; his knowledge of fossils led him to understand that there must have been vast changes in the crust of the earth; he had a remarkable knowledge of anatomy and seems to have foreshadowed Harvey's discovery of the circulation of the blood a century later; he clearly understood the principle of inertia which Galileo was to demonstrate; he even found time to speculate about ships which could move under water and machines in which one day men might be able to fly like birds. 'As iron rusts when it is not used and water gets foul from standing or turns to ice when exposed to cold, so the intellect degenerates without exercise.' Leonardo's mind was in no such danger. Look at the superlative self-portrait, on the back cover, which he drew shortly before going to France: something of the grandeur of this ceaseless searcher for truth comes down to us through over four centuries.

THE BREAK WITH THE PAST BEGINS

The most spectacular scientific advances during the Renaissance were made in the realm of astronomy, but important preliminary work was being done in other fields. For the sake of the patients no branch of knowledge more urgently needed to be rescued from mediaeval practice than medicine. In this task we meet the extraordinary Swiss doctor, Theophrastus Bombast von Hohenheim (c. 1490–1541), who when teaching in Basle took the name of Paracelsus. Arrogant, boastful and sharp-tongued, at his first public lecture he burned the works of Galen in a brass pan with sulphur and proclaimed his superiority over all physicians of the past. Prior to settling in Basle he had worked among the miners of the Tyrol, noting the conditions under which they lived and the accidents and diseases to which they were subject. From 1514 to 1526 he had wandered over much of Europe in order to study the various remedies favoured by different nations. Paracelsus was no great scientist, but his studies had value in that he was led to see the need of applying chemistry to medicine. He stumbled upon what he called 'extract of vitriol', which is clearly

Paracelsus – a painting by Rubens

> **THE BREAK WITH THE PAST BEGINS**

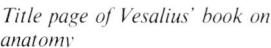

Title page of Vesalius' book on anatomy

ether and which, he said, 'possesses an agreeable taste; even chickens will eat it, whereupon they sleep for a moderately long time, and reawake without having been injured'. It is luckily unknown how many of his patients died from his prescriptions, but his teaching at least directed chemists away from forlorn attempts at making gold to the serious study of medical drugs. Nothing shows his essentially mediaeval outlook better than his contempt for anatomy, with the result that it was not Paracelsus but a Flemish doctor, Andreas Vesalius (1515–64), a scientific student of anatomy, who laid the foundations of modern medicine.

25

> THE BREAK
> WITH THE
> PAST BEGINS

The use of vegetable drugs in the treatment of disease led to a revived interest in the study of plants. Mediaeval monasteries had originally been centres of plant study and there had been no change in the age-old belief that the shape of the leaf and the colour of the flower were signs of the use designed for the plant by its Creator. Greater wealth and security combined with a developing artistic sense led in the Renaissance to the laying out of private parks and gardens and to the creation of special botanic gardens, the first of which was established at Padua in 1545. Herbals, or books containing descriptions of plants together with their use in

Potatoes of Virginia – from a sixteenth-century herbal

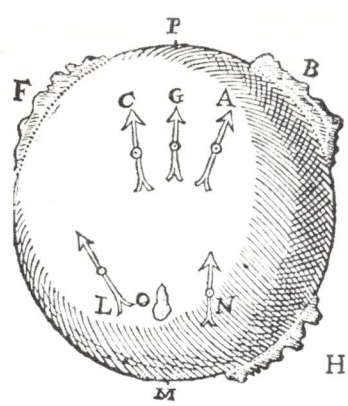

A terrella, or globe-shaped magnet with lumps of iron to represent mountains, and showing the north-seeking property of a magnetic needle – from Gilbert's De Magnete

medicine and cookery, began to appear. The two most famous English examples were those written by William Turner and John Gerard in the second half of the sixteenth century.

One of the marks of a changing approach to science was a slow but increasing reluctance to take the writings of antiquity as gospel truth and to rely on direct experiment. No man better exemplifies this new approach than William Gilbert of Colchester (1540–1603), Fellow of St John's College, Cambridge, and Physician to Queen Elizabeth I. His Cambridge studies were in mathematics, but he soon changed to medicine and moved to London where his house was the centre of scientific discussion and experiment, chiefly in electricity and magnetism. In 1600 he published his book, *De Magnete*, in which he collected all the information previously known on the subject and added the results of his own work. The magnetic needle was known to sailors from the thirteenth century, but no one knew *why* a compass always points in one direction or why a suspended magnetic needle dips downward at an angle to the horizontal. In his book Gilbert maintained that the earth itself must be a magnet and he described experiments to show how the angle of dip varied over the earth's surface. He also studied static electricity and examined the forces developed when certain bodies, such as amber, are rubbed. In fact he coined the name electricity from the Greek word for amber, *elektron*. Gilbert's unbridled scorn was reserved for those who blindly accepted the opinions of others and in this spirit he dedicated his book to those 'who look for knowledge not in books but in things themselves'.

THE COPERNICAN SYSTEM

When Alphonso the Wise of Castile studied the Ptolemaic system, he remarked 'if the Almighty had consulted me before the Creation, I should have recommended something simpler'. What was this system which had been accepted for 1,400 years?

The Pythagorean philosophers about 500 BC had suggested that the earth was spherical and fixed at the centre of a universe of spheres which revolved around it with perfect regularity. The outermost sphere carried the stars; to the other spheres were attached the sun, the moon and the planets. The spheres were invisible and their motions could be deduced only by observing the movement of the heavenly bodies attached to them. The spheres made such perfect music as they revolved that human ears were unfit to hear it. Unfortunately the model did not explain certain strange irregularities in the motion of the planets. The path of Jupiter, for example, appeared to move in a series of loops superimposed on its motion across the heavens.

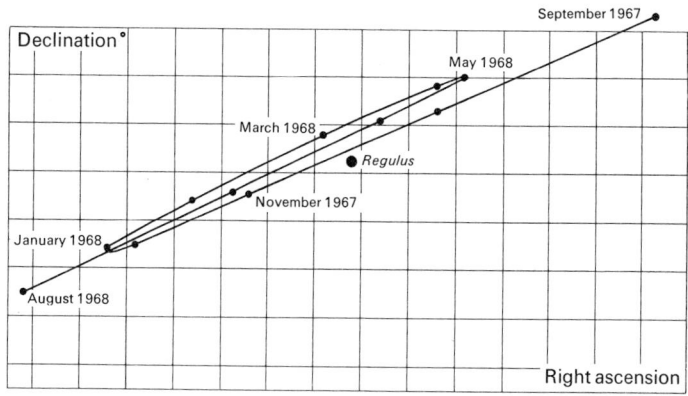

A more complex system of spheres was suggested by Eudoxus about 370 BC. Several spheres were necessary for each planet: the outermost rotated once in every twenty-four hours, the next once in the planet's 'year', and the motions of the inner spheres produced the loops.

Since nothing created by God could be other than perfect, and since the perfect form of movement around

THE COPERNICAN SYSTEM

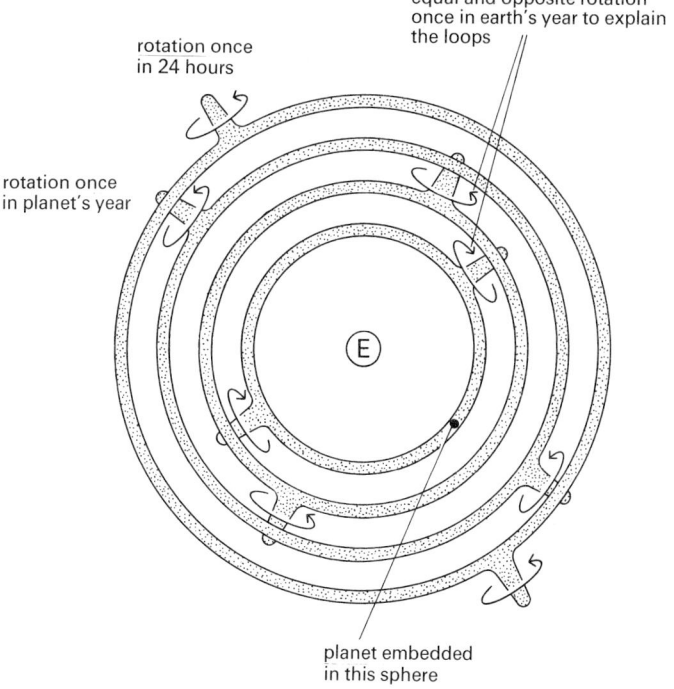

rotation once in 24 hours

rotation once in planet's year

equal and opposite rotation once in earth's year to explain the loops

planet embedded in this sphere

Copernicus

the earth could only be a circle, it became necessary to believe in as many as eighty invisible spheres to account for all the movements which could be observed.

Other models were suggested to account for the loops, including an epicyclic system in which a planet moved in a circle round a circle. This model was developed into an elaborate computing machine by Ptolemy, about AD 120, and it was his ideas which came to be accepted so devoutly in the Middle Ages. 'The astronomer must try his utmost', Ptolemy wrote, 'to explain celestial motions by the simplest possible hypotheses; but if he fails to do so, he must choose whatever hypotheses meet the case.' Unfortunately Ptolemy failed to be simple, as the diagram shows.

It is small wonder that Alphonso the Wise complained of Ptolemy's complexity, for it likewise offended the mathematical mind of a very remarkable man, Nicolaus Koppernigk, or Copernicus in the Latinised version of his name. Copernicus was born in Poland in

> **THE COPERNICAN SYSTEM**

1473. After a lengthy education, both in his own country and in Italy, he returned to Poland in 1506 to take up his duties as Canon of Frauenburg Cathedral, where he remained until his death in 1543. All his life he remained a convinced churchman and an active statesman; he was endowed with a Renaissance breadth of interests, of which mathematics was the chief.

Copernicus was not the first man to consider it possible that the earth moved and that the sun was fixed. Some of the ancient Greek astronomers had advanced this as a theory, notably Aristarchos of Samos, who lived early in the fourth century AD. In the fourteenth century Bishop Nicole Oreme showed that the daily rotation of the earth was at least a possible idea. A century later Nicholas of Cusa made the same claim and his work was probably known to Copernicus. But these were only speculations, unsupported by any proof. What Copernicus did was to postulate the dual

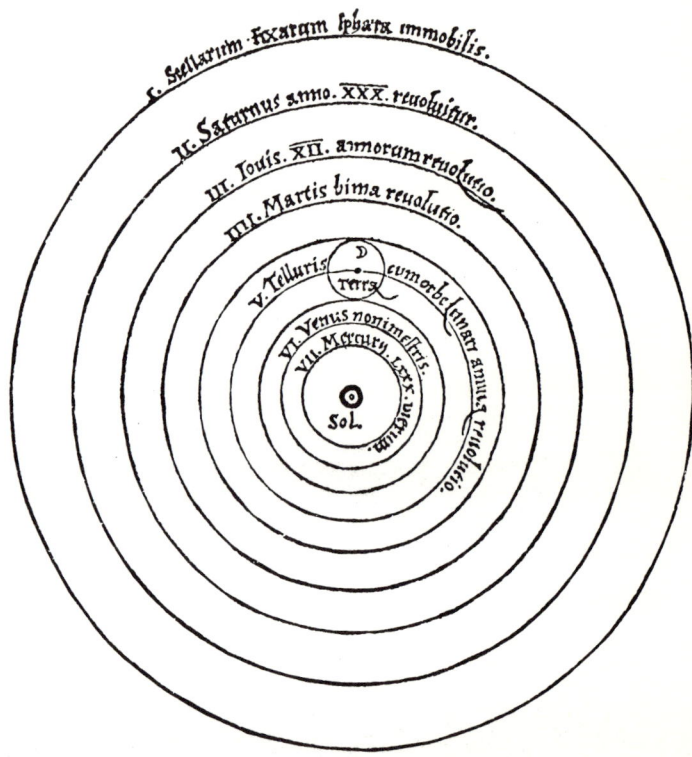

The first diagram of the Copernican system – from Copernicus' book

THE COPERNICAN SYSTEM

motion of the earth, both on its own axis and also round the sun, and then to spend over a quarter of a century of brilliant mathematical calculations proving that his belief was not only possible, but that from it a much more convincing picture of the heavens could be obtained. He accepted what most people would have regarded as an utterly fantastic idea and then with complete honesty and gigantic labour worked out a mathematical system based on it. The triumph of his life was to find that this system accounted for the facts.

For example, the apparent loops in the motion of Jupiter become immediately and easily explicable. With both the earth and Jupiter moving round the sun, the relative direction of Jupiter, seen against the background of stars, changes, giving the looped motion shown above.

Copernicus was an intensely shy man and this is the likeliest reason for the delay in publishing his great book *De Revolutionibus Orbium Caelestium* (*On the Revolutions of the Heavenly Spheres* – not *bodies* but *spheres*). He was not afraid of persecution, for the early stages of his work were well known to the Pope and leading figures in the Church. He shrank from ridicule and thus only divulged his ideas to those able to understand them. He is said to have received the first copy of the book on his death-bed. The title of the book is interesting in that it shows that Copernicus still believed in the existence of the spheres, which he had in his system reduced to thirty-four. To this extent he was wrong, and he died believing that the greatly simplified system which he had worked out would even more firmly establish the majesty of God's creation in the minds of all men. 'We find therefore', he wrote, 'under the orderly arrangement a marvellous symmetry in the universe and a definite relation of harmony in the motion and magnitude of the orbs: of a kind which cannot be obtained in any other way.' These are the words of a man who was not only a mathematician but also a poet and a mystic.

At first sight it might appear that Copernicus had

THE COPERNICAN SYSTEM

not really upset the old astronomical machinery completely. True, he had put the sun and not the earth at the centre, but the general organization remained apparently unchanged, though greatly simplified. But if eighty spheres could be reduced to thirty-four, might not later thinkers discard them altogether? In fact, astronomers and mathematicians in the century after Copernicus's death completed the work which he had begun and which rightly bears his name. It steadily became clear that the publication of his book in 1543 began the destruction of the entire Ptolemaic system and ended the mediaeval picture of man's relation to the universe. For mediaeval man had been taught to believe in the complete distinction between celestial matter and motion on the one hand and, on the other, everything which in his words existed 'beneath the moon'. Thus, gone was the idea of heaven as a place above the skies to which the souls of the faithful were sent.

Far more upsetting was man's new position in the cosmic scheme of things. In the past he had been at the centre of the universe; now his place was uncertain and a subject for endless speculation. By the middle of the seventeenth century astrology, the belief that men's careers and temperaments are 'influenced' by the stars, had fallen from a science to a superstition – though one which even today still lingers in popular newspapers, as does the word 'influenza' whose origin is easy to discern.

THE COUNTER-REFORMATION

Copernicus had dedicated his book to the Pope and his retiring nature made it highly improbable that his theories would meet with official disapproval. Moreover, much of his outlook remained mediaeval and the revolutionary consequences of his book still lay in the future. However, the toleration shown to men of learning in the early Renaissance was ending. The Reformation demanded by Luther and Calvin in the first half of the sixteenth century led to the Counter-Reformation in the second half. The Church struck back in four ways. Between 1545 and 1563 the Council of Trent reaffirmed traditional doctrine; Ignatius Loyola founded the Jesuits, who became the spearhead of the Church's attack; the Index, or list of books forbidden to be read by good Catholics, was first published in

The Council of Trent – painting by Titian

THE COUNTER-REFORMATION

1559; the Inquisition was revived. It was not surprising, therefore, that those who gave prominence to the Copernican system, and who developed it, might do so at considerable risk.

Giordano Bruno (1548–1600), a distinguished Italian philosopher, was the first victim of the new intolerance. He insisted that Copernican ideas were essential to the understanding of a universe which, he taught, was infinitely vast and without any centre, and in which the innumerable stars were suns. It was unlikely that he would have been severely punished for his attack upon traditional scientific views; his crime was to combine with it an attack upon religious mysteries in which he went far beyond Luther. He was forced to take refuge in the republic of Venice, the most enlightened Italian state, where he continued to write and teach. Unwisely Bruno left Venice in 1593, was arrested, tried as a heretic in 1594 and imprisoned. His refusal to recant his beliefs led to his death by burning in 1600. 'I have fought, that is much; victory is in the hands of fate. Be that as it may, this at least future ages will not deny me, be the victor who he may – that I did not fear to die. I yielded to none of my fellows in constancy and preferred a spirited death to a cowardly life.' Such a spirit no Inquisition could permanently destroy.

TYCHO BRAHE AND KEPLER

More important to the final acceptance of the Copernican system was the work of two men, for a brief period in partnership.

Tycho Brahe (1546–1601) was a Dane of noble birth to whom, in 1576, King Frederick II offered the island of Hveen near Elsinore, on which to build an observatory. Attached to the observatory were a library, a printing-shop, a chemical laboratory and a workshop where his instruments could be made; the whole was set in pleasant gardens and christened by Tycho Brahe, Uraniborg, the Tower of Heaven. Tycho Brahe was never a complete upholder of Copernicus, for he could not bring himself to believe that this 'heavy, sluggish' earth could move. In fact, he adopted a compromise system of his own in which the planets revolved around the sun, and together sun and planets revolved around the earth.

Tycho Brahe

His theories do not matter: his brilliant observations made nightly over twenty years constitute Tycho Brahe's claim to fame. He was the first to appreciate that haphazard observations over a short period were inadequate and that what was needed were precise observations over many years. For this purpose he

TYCHO BRAHE AND KEPLER

designed and constructed a whole series of instruments. He had early realised that there were discrepancies of up to five degrees between the results of the observations and the calculations of Ptolemy and Copernicus. He set himself, without the aid of a telescope, which did not yet exist, to produce a far more accurate set of figures and he succeeded.

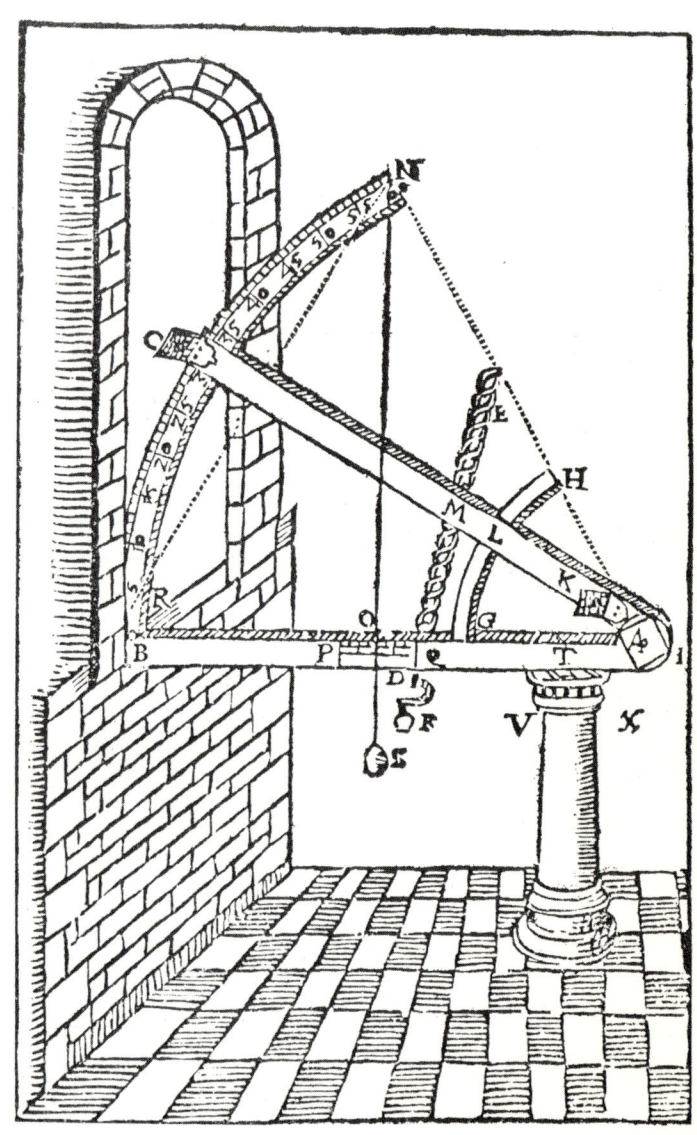

Tycho Brahe's quadrant

TYCHO BRAHE AND KEPLER

In the course of his work he made two important discoveries. In November 1572 a new star appeared in the constellation Cassiopeia. According to Greek science no change was possible among the heavenly bodies above the sphere of the moon. Tycho Brahe tried to measure its parallax, that is the shift in position which should be observable when seen from opposite sides of the earth's orbit. He could detect none, and so concluded that a new star existed and that the stars were subject to change. He made his second discovery five years later in 1577 when a comet appeared. Hitherto it had been thought that comets were simply exhalations of the earth's atmosphere. By measuring its parallax Tycho Brahe proved that it must be much more distant from the earth than the moon was and therefore that comets must be heavenly bodies.

Tycho Brahe was an arrogant and ill-tempered man, who easily made enemies. As a result, with a new king on the Danish throne, he was forced to leave Uraniborg in 1597 and to accept the Emperor Rudolf II's offer of a castle near Prague for his observatory. Here he was joined for the few remaining years of his life by a German assistant, Johann Kepler (1571–1630). Tycho Brahe had been a wonderful observer but a poor mathematician; Kepler was a brilliant mathematician who could never be much of an observer since his eyesight had been seriously injured by early illness. Although they worked together for less than four years Kepler was able to take over and complete the work of Brahe.

Kepler's life, except for the brief partnership with Tycho Brahe, was a long story of acute financial and domestic worry made harder to bear by constant ill-health. He is historically interesting as a man whose discoveries about how planets moved prepared the way for Newton and who at the same time was a believer in astrology. Much in the twenty volumes of his published writings is nonsense; but in 1627 he discharged his debt to Tycho Brahe by publishing his completed calculations, a set of tables indispensable to seventeenth-century navigators. Kepler was always convinced that

Kepler

TYCHO BRAHE AND KEPLER

Frontispiece of Kepler's astronomical tables, dedicated to the Emperor Rudolf II

God created the world in accordance with the principle of perfect numbers and he shook off old beliefs with extreme reluctance. He hoped to prove mathematically that heavenly bodies moved in a perfect circle but, being a man of complete intellectual integrity, he was soon forced to consider first an oval path and finally an ellipse. Mathematics could not be gainsaid, though he referred to his discovery as 'a cartful of dung'. His labours were intense. 'If you find this work difficult

TYCHO BRAHE AND KEPLER

and wearisome to follow,' he wrote, 'take pity on me, for I have repeated these calculations seventy times, nor be surprised that I have spent five years on this theory of Mars.' At the time he was Professor of Mathematics at Linz University, supplementing his income by telling fortunes and publishing a kind of Old Moore's Almanack, while also trying, with statistical thoroughness, to choose a second wife from eleven candidates!

From studying the orbit of Mars Kepler turned to study the alteration in the speed of motion of the planet as it travelled along an elliptical path. From all these calculations he enunciated his first two great laws:

1 The planet describes an ellipse, the sun being at one focus (a word first used by Kepler).

2 A straight line joining the planet to the sun sweeps out equal areas in any two equal intervals of time.

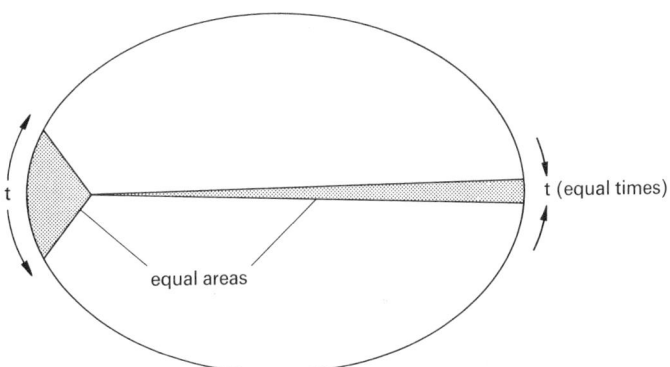

Remembering Kepler's mediaeval belief that the universe had been created according to a geometrical plan, it is not surprising that he sought for a relationship between the various measurements of the solar system, thus discovering his third law:

3 The cubes of the mean distances (R) of the planets from the sun bear a constant ratio to the squares of their times of revolution (T) about the sun. (See the following table.)

TYCHO BRAHE AND KEPLER

Planet	Mean distance (R) from sun in kilometres	Period of revolution (T) in days	R^3/T^2
Mercury	0.58×10^8	88.0	2.506×10^{19}
Venus	1.08×10^8	224.7	2.496×10^{19}
Earth	1.49×10^8	365.3	2.497×10^{19}
Mars	2.28×10^8	687.0	2.511×10^{19}
Jupiter	7.78×10^8	4 333.0	2.509×10^{19}
Saturn	14.3×10^8	10 760.0	2.525×10^{19}

What was the cause of such motion Kepler never knew and he was content to believe in the work of angels or the agency of some special genii. Although he was a Protestant his work was condemned as early as 1596 by the University of Tübingen on the same grounds as those on which the Inquisition was to condemn Galileo nearly forty years later. Neither Catholic nor Protestant had the monopoly of intolerance. But Kepler's condemnation was not widespread in its effect and his work continued till his death. Basically his life was devoted to searching for ultimate causes, for the mathematical harmonies in the mind of the Creator: in this search it happened that he made acceptance of the Copernican system in the long run inevitable.

GALILEO

With the career of Galileo Galilei (1564–1642) we reach the threshold of the modern world. Not only, thanks largely to the invention of the telescope, did he finally overthrow the Ptolemaic system and prove Copernicus to have been fundamentally correct, but the combination of experimental and mechanical skill allied to mathematical genius enabled him to become the founder of the science of dynamics. This latter part of his work was more important than his astronomy, for it prepared the way for Newton's discoveries. Furthermore, Galileo's entire approach to science was modern. The mediaeval Schoolmen had taught that the whole of nature was a unified and logical system created for the benefit of man. To Galileo and to all future scientists the secrets of nature were waiting to be unravelled; henceforward each fact acquired by observation or experiment must be accepted as it stood, regardless of the consequences to any existing beliefs.

Galileo was born at Pisa in the same year as Shakespeare. At the age of seventeen he went to the university at Pisa and nine years later found himself Professor of Mathematics there, although his original subject was medicine. In all universities the doctrines of Aristotle on motion were taught and accepted without question. Bodies were regarded as being either heavy or light and as falling or rising with a velocity proportional to their heaviness or lightness. For example, a ten-kilogramme shot was believed to fall ten times as fast as a one-kilogramme shot. In the teaching of science there was then no practical work and boys at school accepted unhesitatingly what Greek scientists had written. Nevertheless, it is curious that until about 1586 nobody apparently ever conducted an experiment to find out whether Aristotle's statements on motion were accurate or not. By that date Stevinus of Bruges had certainly shown that the two shot would have reached the ground simultaneously and Galileo probably repeated the experiment, though not from the top of the Leaning Tower of Pisa as legend asserts.

Another picturesque but probably false story tells of Galileo's discovery of the properties of the pendulum.

Galileo

GALILEO

It is said that at the age of eighteen in the cathedral at Pisa his attention was distracted from his prayers by a lamp swinging overhead. He timed its swings by the beats of his own pulse – no watch had yet been invented – and found that the time of a single swing apparently remained the same whether the lamp swung through a wider arc or a narrower one. Later more accurate experiments confirmed this.

It is not possible to enter into a description of all Galileo's experiments. Basically he set out to discover not *why* bodies fall but *how*. Setting a ball rolling down an inclined plane and timing it with a water clock of his own devising, he calculated that the distance travelled by the ball varied as the square of the time of descent. Further experiments proved that the path of a projectile was a parabola, a discovery of practical importance in gunnery. More important was his enunciation of the principle of inertia. Hitherto it had always been assumed that every motion on earth needed a continual force to maintain it. In the heavens Aristotle's Unmoved Mover kept the planets going in circles. What Galileo discovered was that it is not motion but the starting or stopping of motion, or the change in its direction, which demands external force. Galileo asserted the tendency of a body to continue in uniform motion in a straight line and to resist a change of motion, a concept of great importance later to Newton.

In 1592 Galileo left Pisa where his outspoken comments on his colleagues had made him very unpopular. He moved to Padua, in the republic of Venice, to become Professor of Mathematics. Here he invented a thermometer, but his contributions to physics were soon overshadowed by his open adherence to the teachings of Copernicus and his proof of their truth. Galileo never suffered fools gladly, but he probably proceeded with some caution in view of the fate of Giordano Bruno. However, the appearance in 1604 of another brilliant new star made it increasingly impossible for those with eyes to see to accept Aristotle's doctrine of the unchanging heavens. Five years later

GALILEO

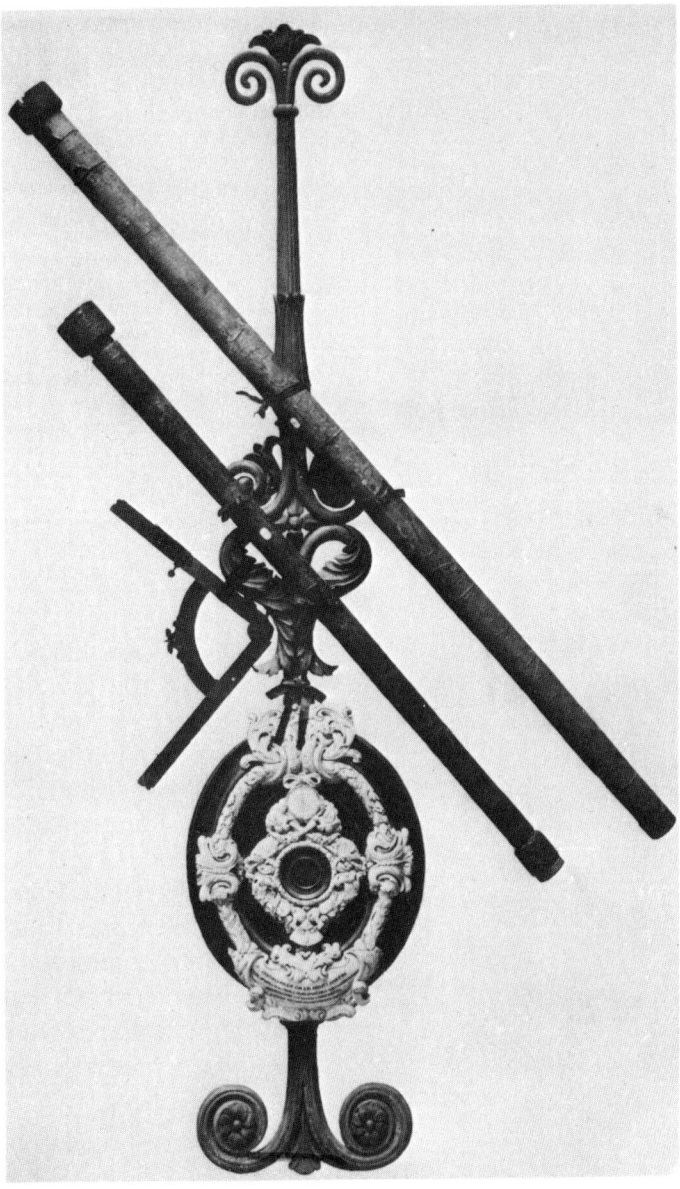

Galileo's original telescope

the whole Ptolemaic system was destroyed, not by mathematics but by the chance invention of the telescope. The most likely inventor was Hans Lippershey, a spectacle maker of Middelburg in Holland, whose apprentice accidentally arranged a combination of lenses which brought distant objects into near vision.

GALILEO

Galileo described his own reaction to the news of this invention:

> About ten months ago a report reached my ears that a Dutchman had constructed a telescope, by the aid of which visible objects, although at a great distance from the eye of the observer, were seen distinctly as if near; and some proofs of its most wonderful performances were reported, which some gave credence to, but others contradicted. A few days after, I received confirmation of the report in a letter written from Paris by a noble Frenchman, which finally determined me to give myself up first to enquire into the principle of the telescope, and then to consider the means by which I might compass the invention of a similar instrument, which after a little while I succeeded in doing, through deep study of the theory of refraction; and I prepared a tube, at first of lead, in the ends of which I fitted two glass lenses, both plane on one side, but on the other side one spherically convex and the other concave. Then, bringing my eye to the concave lens, I saw objects satisfactorily large and near, for they appeared one third of the distance off and nine times larger than when they are seen with the natural eye alone. I shortly afterward constructed another telescope with more nicety, which magnified objects more than sixty times. At length, by sparing neither labour nor expense, I succeeded in constructing for myself an instrument so superior that objects seen through it appear magnified nearly a thousand times, and more than thirty times nearer than if viewed by the natural powers of sight alone.

What Galileo did not say was that his telescope was better than the Dutch model which had the distinct disadvantage of giving an inverted image. Galileo presented one to the Doge, and Venetian senators were thrilled to climb the Campanile of St Mark's to look at their fleet far out at sea but appearing miraculously close at hand. Galileo's salary was doubled.

However, it was when Galileo turned his increasingly efficient telescope on the heavens that trouble threatened. He was amazed at the multitude of new stars now visible and he realised that the Milky Way was 'nothing else but a mass of innumerable stars planted together in clusters'. He saw detail on the moon clearly for the first time:

> Let me first speak of the surface of the moon, which is turned toward us. For the sake of being understood more easily I distinguish two parts in it, which I call respectively the brighter and the darker. The brighter part seems to surround and pervade the whole hemisphere; but the darker part, like a sort of cloud, discolours the moon's surface and makes it appear covered with spots. Now these spots, as they are somewhat dark and of considerable size, are plain to everyone, and every age has seen them, wherefore I shall call them *great* or *ancient* spots, to distinguish them from other spots, smaller in size, but so thickly scattered that they sprinkle the whole surface of the moon, but especially the brighter portion of it. These spots have never been observed by anyone before me; and from my

GALILEO

observations of them, often repeated, I have been led to that opinion which I have expressed, namely, that I feel sure that the surface of the moon is not perfectly smooth, free from inequalities, and exactly spherical, as a large school of philosophers considers with regard to the moon and the other heavenly bodies, but that, on the contrary, it is full of inequalities, uneven, full of hollows and protuberances, just like the surface of the earth itself, which is varied everywhere by lofty mountains and deep valleys.

The appearances from which we may gather these conclusions are of the following nature. On the fourth or fifth day after new moon, when the moon presents itself to us with bright horns, the boundary which divides the part in shadow from the enlightened part does not extend continuously in an ellipse, as would happen in the case of a perfectly spherical body, but it is marked out by an irregular, uneven and very wavy line... for several bright excrescences, as they may be called, extend beyond the boundary of light and shadow into the dark part, and on the other hand pieces of shadow encroach upon the light – nay, even a great quantity of small blackish spots, altogether separated from the dark part, sprinkle everywhere almost the whole space which is at the time flooded with the sun's light, with the exception of that part alone which is occupied by the great and ancient spots. I have noticed that the small spots just mentioned have this common characteristic always and in every case, that they have the dark part towards the sun's position, and on the side away from the sun they have brighter boundaries, as if they were crowned with shining summits.

Now we have an appearance quite similar on the earth about sunrise, when we behold the valleys, not yet flooded with light, but the mountains surrounding them on the side opposite to the sun already ablaze with the splendour of his beams; and just as the shadows in the hollows of the earth diminish in size as the sun rises higher, so also these spots on the moon lose their blackness as the illuminated part grows larger and larger. Again, not only are the boundaries of light and shadow in the moon seen to be uneven and sinuous, but – and this produces still greater astonishment – there appear very many bright points within the darkened portion of the moon, altogether divided and broken off from the illuminated tract, and separated from it by no inconsiderable interval, which after a little while gradually increase in size and brightness, and after an hour or two become joined onto the rest of the main portion, now become somewhat larger; but in the meantime others, one here and another there, shooting up as if growing, are lighted up within the shaded portion, increase in size, and at last are linked onto the same luminous surface, now still more extended...

Now is it not the case on the earth before sunrise that, while the level plain is still in shadow, the peaks of the most lofty mountains are illuminated by the sun's rays? After a little while does not the light spread further, while the middle and larger parts of those mountains are becoming illuminated; and at length, when the sun has risen, do not the illuminated parts of the plains and hills join together? The grandeur, however, of such prominences and depressions in the moon seems to surpass both in magnitude and extent the ruggedness of the earth's surface, as I shall hereafter show...

Furthermore Galileo noticed spots on the sun: what a blemish in a supposedly perfect creation! Most exciting of all was his study of Jupiter, the story of which he told in his *Siderius Nuncius (The Starry Messenger)*. An hour after sunset on the night of 7 January 1610 he looked through his telescope and noticed three fixed

GALILEO

stars near the planet Jupiter, two on one side and one on the other. He had been finding so many new stars that he thought little of this. By pure chance next night he looked again at Jupiter and this time saw all three stars on the same side to the west. Thinking that Jupiter must be behaving in an extraordinary way he waited impatiently for the next night, only to be frustrated by thick cloud. Eventually on 11 January he saw only two stars, both to the east of the planet, and guessed what this meant. Two nights later his guess was confirmed when he saw all four satellites of Jupiter and realised that, like the earth, Jupiter had a moon, or rather moons. What in effect he was seeing was the Copernican system in miniature.

In 1610 Galileo accepted an invitation from the Grand Duke of Tuscany to return to his native state in a position which enabled him to give up lecturing and devote the whole of his time to his researches. It was a fateful decision, for it meant leaving the safety of Venice for territory where the power of the Church was great. Nevertheless, he began the study of hydrostatics in Florence, observed apparent 'satellites' of Saturn, subsequently identified by Huyghens as rings, and wrote about how to determine longitude. He never ceased to attack the Aristotelian teaching, with the result that in 1616, under threat of torture and imprisonment, he was ordered 'to abandon and cease to teach his false, impious, and heretical opinions'.

The years which followed were at first comparatively quiet. Much of his work did not bring him into conflict with the Church, and when in 1623 his great friend and supporter became Pope Urban VIII Galileo hoped that a new era of tolerance would begin. He thus felt safe in writing the great book of his life, *The Dialogues concerning the Two Chief Systems of the World,* which he finished in 1630. In order to remain technically true to the promise extracted from him in 1616 the book took the form of a discussion between Salviati, a Copernican, Simplicio, an Aristotelian, and Sagredo, who acted as chairman. There was little doubt that Salviati got the better of the argument and, when

GALILEO

Title page from Galileo's Dialogues

Galileo's enemies persuaded Pope Urban that Simplicio was intended as a caricature of himself, Galileo was summoned to Rome for trial before the Inquisition. He was now an old man and his health was beginning to fail; he was inevitably reminded by his friends of the fate of Giordano Bruno and urged to submit. In June 1633, broken in spirit, Galileo made the famous recantation of his beliefs in a document later read from every pulpit and in every university. It was a tragic moment. Although his books were banned, Galileo was allowed to live in strict retirement near Florence where

47

GALILEO

friends could come and visit him. His intellect was unimpaired and he continued his work on dynamics, publishing his conclusions in his *Dialogues on Motion*, printed in 1636 in Amsterdam. However, his health continued to deteriorate and he lost both sight and hearing. In January 1642 death came as a merciful release. On the following Christmas Day Isaac Newton was born – an as yet unrealised guarantee that Galileo's work would continue.

'When I consider what marvellous things and how many of them men have understood, inquired into and contrived, I recognise and understand only too clearly that the human mind is a work of God's, and one of the most excellent.' So wrote Galileo, and his words can be used as a fitting epitaph on the work of all the Renaissance scientists who strove in their varying ways to break free from the shackles of the past, and thus to make possible the true beginnings of modern science in the seventeenth century.